tredition®

tredition was established in 2006 by Sandra Latusseck and Soenke Schulz. Based in Hamburg, Germany, tredition offers publishing solutions to authors and publishing houses, combined with worldwide distribution of printed and digital book content. tredition is uniquely positioned to enable authors and publishing houses to create books on their own terms and without conventional manufacturing risks.

For more information please visit: www.tredition.com

TREDITION CLASSICS

This book is part of the TREDITION CLASSICS series. The creators of this series are united by passion for literature and driven by the intention of making all public domain books available in printed format again - worldwide. Most TREDITION CLASSICS titles have been out of print and off the bookstore shelves for decades. At tredition we believe that a great book never goes out of style and that its value is eternal. Several mostly non-profit literature projects provide content to tredition. To support their good work, tredition donates a portion of the proceeds from each sold copy. As a reader of a TREDITION CLASSICS book, you support our mission to save many of the amazing works of world literature from oblivion. See all available books at www.tredition.com.

Project Gutenberg

The content for this book has been graciously provided by Project Gutenberg. Project Gutenberg is a non-profit organization founded by Michael Hart in 1971 at the University of Illinois. The mission of Project Gutenberg is simple: To encourage the creation and distribution of eBooks. Project Gutenberg is the first and largest collection of public domain eBooks.

Birds in the Calendar

Frederick G. (Frederick George) Aflalo

Imprint

This book is part of TREDITION CLASSICS

Author: Frederick G. (Frederick George) Aflalo
Cover design: Buchgut, Berlin – Germany

Publisher: tredition GmbH, Hamburg - Germany
ISBN: 978-3-8472-1321-5

www.tredition.com
www.tredition.de

Copyright:
The content of this book is sourced from the public domain.

The intention of the TREDITION CLASSICS series is to make world literature in the public domain available in printed format. Literary enthusiasts and organizations, such as Project Gutenberg, worldwide have scanned and digitally edited the original texts. tredition has subsequently formatted and redesigned the content into a modern reading layout. Therefore, we cannot guarantee the exact reproduction of the original format of a particular historic edition. Please also note that no modifications have been made to the spelling, therefore it may differ from the orthography used today.

BIRDS IN THE CALENDAR

BY F. G. AFLALO

LONDON: MARTIN SECKER
NUMBER FIVE JOHN STREET ADELPHI
First Published 1914

NOTE

These sketches of birds, each appropriate to one month of the twelve, originally appeared in *The Outlook*, to the Editor and Proprietors of which review I am indebted for permission to reprint them in book form.

F. G. A.

Easter, 1914.

JANUARY
THE PHEASANT

[11]

THE PHEASANT

As birds are to be considered throughout these pages from any standpoint but that of sport, much that is of interest in connection with a bird essentially the sportsman's must necessarily be omitted. At the same time, although this gorgeous creature, the chief attraction of social gatherings throughout the winter months, appeals chiefly to the men who shoot and eat it, it is not uninteresting to the naturalist with opportunities for studying its habits under conditions more favourable than those encountered when in pursuit of it with a gun.

In the first place, with the probable exception of the swan, of which something is said on a later page, the pheasant stands alone among the birds of our woodlands in its personal interest for the historian. It is not, in fact, a British bird, save by acclimatisation, at all, and is generally regarded as a legacy of the Romans. The time and manner of its introduction into Britain are, it is true, veiled in obscurity. What we know, [12] on authentic evidence, is that the bird was officially recognised in the reign of Harold, and that it had already come under the ægis of the game laws in that of Henry I, during the first year of which the Abbot of Amesbury held a licence to kill it, though how he contrived this without a gun is not set forth in detail. Probably it was first treed with the aid of dogs and then shot with bow and arrow. The original pheasant brought over by the Romans, or by whomsoever may have been responsible for its naturalisation on English soil, was a dark-coloured bird and not the type more familiar nowadays since its frequent crosses with other species from the Far East, as well as with several ornamental types of yet more recent introduction.

In tabooing the standpoint of sport, wherever possible, from these chapters, occasional reference, where it overlaps the interests of the field-naturalist, is inevitable. Thus there are two matters in which both classes are equally concerned when considering the pheasant. The first is the real or alleged incompatibility of pheasants and foxes in the same wood. The question of [13] rivalry between pheasant and fox, or (as I rather suspect) between those who shoot the one and hunt the other, admits of only one answer. The fox eats the pheasant; the pheasant is eaten by the fox. This not very complex proposition may read like an excerpt from a French grammar, but it is the epitome of the whole argument. It is just possible—we have no actual evidence to go on—that under such wholly natural conditions as survive nowhere in rural England the two might flourish side by side, the fox taking occasional toll of its agreeably flavoured neighbours, and the latter, we may suppose, their wits sharpened by adversity, gradually devising means of keeping out of the robber's reach. In the artificial environment of a hunting or shooting country, however, the fox will always prove too much for a bird dulled by much protection, and the only possible *modus vivendi* between those concerned must rest on a policy of give and take that deliberately ignores the facts of the case.

More interesting, on academic grounds at any rate, is the process of education noticeable in pheasants in parts of the country [14] where they are regularly shot. Sport is a great educator. Foxes certainly, and hares probably, run the faster for being hunted. Indeed the fox appears to have acquired its pace solely as the result of the chase, since it does not figure in the Bible as a swift creature. The genuine wild pheasant in its native region, a little beyond the Caucasus, is in all probability a very different bird from its half-domesticated kinsman in Britain. I have been close to its birthplace, but never even saw a pheasant there. We are told, on what ground I have been unable to trace, that the polygamous habit in these birds is a product of artificial environment; but what is even more likely is that the true wild pheasant of Western Asia (and not the acclimatised bird so-called in this country) trusts much less to its legs than our birds, which have long since learnt that there is safety in running. Moreover, though it probably takes wing more readily, it is doubtful whether it flies as fast as the pace, something a little short of forty miles an hour, that has been estimated as a common performance in driven birds at home. [15]

The pheasant is in many respects a very curious bird. At the threshold of life, it exhibits, in common with some of its near relations, a precocity very unusual in its class; and the readiness with which pheasant chicks, only just out of the egg, run about and forage for themselves, is astonishing to those unused to it. Another interesting feature about pheasants is the extraordinary difference in plumage between the sexes, a gap equalled only between the blackcock and greyhen and quite unknown in the partridge, quail and grouse. Yet every now and again, as if resentful of this inequality of wardrobe, an old hen pheasant will assume male plumage, and this epicene raiment indicates barrenness. Ungallant feminists have been known to cite the case of the "mule" pheasant as pointing a moral for the females of a more highly organised animal.

The question of the pheasant's natural diet, more particularly where this is not liberally supplemented from artificial sources, brings the sportsman in conflict with the farmer, and a demagogue whose zeal occasionally outruns his discretion has even [16] endeavoured to cite the mangold as its staple food. This, however, is political, and not natural history. Although, however, like all grain-eating birds, the pheasant is no doubt capable of inflicting appreciable damage on cultivated land, it seems to be established beyond all question that it also feeds greedily on the even more destructive larva of the crane-fly, in which case it may more than pay its footing in the fields. The foodstuff most fatal to itself is the yew leaf, for which, often with fatal results, it seems to have an unconquerable craving. The worst disease, however, from which the pheasant suffers is "gapes," caused by an accumulation of small red worms in the windpipe that all but suffocate the victim.

Reference has been made to the bird's great speed in the air, as well as to its efficiency as a runner. It remains only to add that it is also a creditable swimmer and has been seen to take to water when escaping from its enemies.

The polygamous habit has been mentioned. Ten or twelve eggs, or more, are laid in the simple nest of leaves, and this is generally [17] placed on the ground, but occasionally in a low tree or hedge, or even in the disused nest of some other bird.

Comparatively few of the birds referred to in the following pages appeal strongly to the epicure, but the pheasant, if not, perhaps, the most esteemed of them, is at least a wholesome table bird. It should, however, always be eaten with chip potatoes and bread sauce, and not in the company of cold lettuce. Those who insist on the English method of serving it should quote the learned Freeman, who, when confronted with the Continental alternative, complained bitterly that he was not a silkworm!

FEBRUARY
THE WOODCOCK

[21]

THE WOODCOCK

There are many reasons why the woodcock should be prized by the winter sportsman more than any other bird in the bag. In the first place, there is its scarcity. Half a dozen to every hundred pheasants would in most parts of the country be considered a proportion at which none could grumble, and there are many days on which not one is either seen or shot. Again, there is the bird's twisting flight, which, particularly inside the covert, makes it anything but an easy target. Third and last, it is better to eat than any other of our wild birds, with the possible exception of the golden plover. Taking one consideration with another, then, it is not surprising that the first warning cry of "Woodcock over!" from the beaters should be the signal for a sharp and somewhat erratic fusillade along the line, a salvo which the beaters themselves usually honour by crouching out of harm's way, since they know from experience that even ordinarily cool and collected shots are [22] sometimes apt to be fired with a sudden zeal to shoot the little bird, which may cost one of them his eyesight. According to the poet,

"Lonely woodcocks haunt the watery glade;"

and so no doubt they do at meal-time after sunset, but we are more used to flushing them amid dry bracken or in the course of some frozen ditch. Quite apart, however, from its exhilarating effect on the sportsman, the bird has quieter interests for the naturalist, since in its food, its breeding habits, its travels, and its appearance it combines more peculiarities than perhaps any other bird, certainly than any other of the sportsman's birds, in these islands. It is not, legally speaking, a game bird and was not included in the Act of 1824, but a game licence is required for shooting it, and it enjoys since 1880 the protection accorded to other wild birds. This is excellent, so far as it goes, but it ought to be protected during the same period as the pheasant, particularly now that it is once more established as a resident species all over Britain and Ireland.

This new epoch in the history of its adventures in these islands is the work of the [23] Wild Birds' Protection Acts. In olden times, when half of Britain was under forest, and when guns were not yet invented that could "shoot flying," woodcocks must have been much more plentiful than they are to-day. In those times the bird was taken on the ground in springes or, when "roding" in the mating season, in nets, known as "shots," that were hung between the trees. When the forest area receded, the resident birds must have dwindled to the verge of extinction, for on more than one occasion we find even a seasoned sportsman like Colonel Hawker worked up to a rare pitch of excitement after shooting woodcock in a part of Hampshire where in our day these birds breed regularly. Thanks, however, to the protection afforded by the law, there is once again probably no county in England in which woodcocks do not nest.

At the same time, it is as an autumn visitor that, with the first of the east wind in October or November, we look for this untiring little traveller from the Continent. Some people are of opinion that since it has extended its residential range fewer come oversea to swell the numbers, but the arrivals [24] are in some years considerable, and if a stricter watch were kept on unlicensed gunners along the foreshore of East Anglia, very much larger numbers would find their way westwards instead of to Leadenhall. As it is, the wanderers arrive, not necessarily, as has been freely asserted, in poor condition, but always tired out by their journey, and numbers are secured before they have time to recover their strength. Yet those which do recover fly right across England, some continuing the journey to Ireland, and stragglers even, with help no doubt from easterly gales, having been known to reach America.

The woodcock is interesting as a parent because it is one of the very few birds that carry their young from place to place, and the only British bird that transports them clasped between her legs. A few others, like the swans and grebes, bear the young ones on the back, but the woodcock's method is unique. Scopoli first drew attention to his own version of the habit in the words *"pullos rostro portat,"* and it was old Gilbert White who, with his usual eye to the practical, doubted whether so long and slender a bill [25] could be turned to such a purpose. More recent observation has confirmed White's objection and has established the fact of the woodcock holding the young one between her thighs, the beak being apparently used to steady her burden. Whether the little ones are habitually carried about in this fashion, or merely on occasion of danger, is not known, and indeed the bird's preference for activity in the dusk has invested accurate observation of its habits with some difficulty. Among well-known sportsmen who were actually so fortunate as to have witnessed this interesting performance, passing mention may be made of the late Duke of Beaufort, the Hon. Grantley Berkeley, and Sir Ralph Payne-Gallwey.

Reference has already been made to the now obsolete use of nets for the capture of these birds when "roding." The cock-shuts, as they were called, were spread so as to do their work after sundown, and this is the meaning of Shakespeare's allusion to "cock-shut time." This "roding" is a curious performance on the part of the males only, and it bears some analogy to the "drumming" of snipe. It is accompanied indeed by the same [26] vibrating noise, which may be produced from the throat as well, but is more probably made only by the beating of the wings. There appears to be some divergence of opinion as to its origin in both birds, though in that of the snipe such sound authorities as Messrs. Abel Chapman and Harting are convinced that it proceeds from the quivering of the primaries, as the large quill-feathers of the wings are called. Other naturalists, however, have preferred to associate it with the spreading tail-feathers. Whether these eccentric gymnastics are performed as displays, with a view to impressing admiring females, or whether they are merely the result of excitement at the pairing season cannot be determined. It is safe to assume that they aim at one or other of these objects, and further no one can go with any certainty. The word "roding" is spelt "roading" by Newton, who thus gives the preference to the Anglo-Saxon description of the aërial tracks followed by the bird, over the alternative derivation from the French "roder," which means to wander. The flight is at any rate wholly different from that to which the sportsman is accustomed when one [27] of these birds is flushed in covert. In the latter case, either instinct or experience seems to have taught it extraordinary tricks of zigzag manœuvring that not seldom save its life from a long line of over-anxious guns; though out in the open, where it generally flies in a straight line for the nearest covert, few birds of its size are easier to bring down. Fortunately, we do not in England shoot the bird in springtime, the season of "roding," but the practice is in vogue in the evening twilight in every Continental country, and large bags are made in this fashion.

In its hungry moments the woodcock, like the snipe, has at once the advantages and handicap of so long a beak. On hard ground, in a long spell of either drought or frost, it must come within measurable distance of starvation, for its only manner of procuring its food in normal surroundings is to thrust its bill deep into the soft mud in search of earthworms. The bird does not, it is true, as was once commonly believed, live by suction, or, as the Irish peasants say in some parts, on water, but such a mistake might well be excused in anyone who had watched the [28] bird's manner of digging for its food in the ooze. The long bill is exceedingly sensitive at the tip, and in all probability, by the aid of a tactile sense more highly developed than any other in our acquaintance, this organ conveys to its owner the whereabouts of worms wriggling silently down out of harm's way. On first reaching Britain, the woodcock remains for a few days on the seashore to recover from its crossing, and at this time of rest it trips over the wet sand, generally in the gloaming, and picks up shrimps and such other soft food as is uncovered between tidal marks. It is not among the easiest of birds to keep for any length of time in captivity, but if due attention be paid to its somewhat difficult requirements in the way of suitable food, success is not unattainable. On the whole, bread and milk has been found the best artificial substitute for its natural diet. With the *kiwi* of New Zealand, a bird not even distantly related to the woodcock, and a cousin rather of the ostrich, but equipped with much the same kind of bill as the subject of these remarks, an even closer imitation of the natural food has been found possible in [29] menageries. The bill of the *kiwi*, which has the nostrils close to the tip, is even more sensitive than that of the woodcock and is employed in very similar fashion. At Regent's Park the keeper supplies the bird with fresh worms so long as the ground is soft enough for spade-work. They are left in a pan, and the *kiwi* eats them during the night. In winter, however, when worms are not only hard to come by in sufficient quantity but also frost-bitten and in poor condition, an efficient substitute is found in shredded fillet steak, which, whether it accepts it for worms or not, the New Zealander devours with the same relish.

When a woodcock lies motionless among dead leaves, it is one of the most striking illustrations of protective colouring to be found anywhere. Time and again the sportsman all but treads on one, which is betrayed only by its large bright eye. There are men who, in their eagerness to add it to the bag, do not hesitate in such circumstances to shoot a woodcock on the ground, but a man so fond of ground game should certainly be refused a game licence and should be allowed to shoot nothing but rabbits.

MARCH
THE WOODPIGEON

[33]

THE WOODPIGEON

The woodpigeon is many things to many men. To the farmer, who has some claim to priority of verdict, it is a curse, even as the rabbit in Australia, the lemming in Norway, or the locust in Algeria. The tiller of the soil, whose business brings him in open competition with the natural appetites of such voracious birds, beasts, or insects, regards his rivals from a standpoint which has no room for sentiment; and the woodpigeons are to our farmers, particularly in the well-wooded districts of the West Country, even as Carthage was to Cato the Censor, something to be destroyed.

It is this attitude of the farmer which makes the woodpigeon pre-eminently the bird of February. All through the shooting season just ended, a high pigeon has proved an irresistible temptation to the guns, whether cleaving the sky above the tree-tops, doubling behind a broad elm, or suddenly swinging out of a gaunt fir. Yet it is in February, when other shooting is at an end and the coverts [34] no longer echo the fusillade of the past four months, that the farmers, furious at the sight of green root-crops grazed as close as by sheep and of young clover dug up over every acre of their tilling, welcome the co-operation of sportsmen glad to use up the balance of their cartridges in organised pigeon battues. These gatherings have, during the past five years, become an annual function in parts of Devonshire and the neighbouring counties, and if the bag is somewhat small in proportion to the guns engaged, a wholesome spirit of sport informs those who take part, and there is a curiously utilitarian atmosphere about the proceedings. Everyone seems conscious that, in place of the usual idle pleasure of the covert-side or among the turnips, he is out for a purpose, not merely killing birds that have been reared to make his holiday, but actually helping the farmers in their fight against Nature. As, moreover, recent scares of an epidemic not unlike diphtheria have precluded the use of the birds for table purposes, the powder is burnt with no thought of the pot.

The usual plan is to divide the guns in [35] small parties and to post these in neighbouring plantations or lining hedges overlooking these spinneys. At a given signal the firing commences and is kept up for several hours, a number of the marauders being killed and the rest so harried that many of them must leave the neighbourhood, only to find a similar warm welcome across the border. Some such concerted attack has of late years been rendered necessary by the great increase in the winter invasion from overseas. It is probable that, as most writers on the subject insist, the wanderings of these birds are for the most part restricted to these islands and are mere food forays, like those which cause locusts to desert a district that they have stripped bare for pastures new. At the same time, it seems to be beyond all doubt the fact that huge flocks of woodpigeons reach our shores annually from Scandinavia, and their inroads have had such serious results that it is only by joint action that their numbers can be kept under. For such work February is obviously the month, not only because most of their damage to the growing crops and seeds is accomplished at [36] this season, but also because large numbers of gunners, no longer able to shoot game, are thus at the disposal of the farmers and only too glad to prolong their shooting for a few weeks to such good purpose.

Many birds are greedy. The cormorant has a higher reputation of the sort to live up to than even the hog, and some of the hornbills, though less familiar, are endowed with Gargantuan appetites. Yet the ringdove could probably vie with any of them. Mr. Harting mentions having found in the crop of one of these birds thirty-three acorns and forty-four beech-nuts, while no fewer than 139 of the latter were taken, together with other food remains, from another. It is no uncommon experience to see the crop of a woodpigeon that is brought down from a great height burst, on reaching the earth, with a report like that of a pistol, and scatter its undigested contents broadcast. Little wonder then, that the farmers welcome the slaughter of so formidable a competitor! It is one of their biggest customers, and pays nothing for their produce. One told me, not long ago, that the woodpigeons had got at a little patch [37] of young rape, only a few acres in all, which had been uncovered by the drifting snow, and had laid it as bare as if the earth had never been planted. Seeing what hearty meals the woodpigeon makes, it is not surprising that it should sometimes throw up pellets of undigested material. This is not, however, a regular habit, as in the case of hawks and owls, and is rather, perhaps, the result of some abnormally irritating food.

Pigeons digest their food with the aid of a secretion in the crop, and it is on this soft material, popularly known as "pigeons' milk," that they feed their nestlings. This method suggests analogy to that of the petrels, which rear their young on fish-oil partly digested after the same fashion. Indeed, all the pigeons are devoted parents. Though the majority build only a very pretentious platform of sticks for the two eggs, they sit very close and feed the young ones untiringly. Some of the pigeons of Australia, indeed, go even further. Not only do they build a much more substantial nest of leafy twigs, but the male bird actually sits throughout the day, such paternal sense of [38] duty being all the more remarkable from the fact that these pigeons of the Antipodes usually lay but a single egg. Australia, with the neighbouring islands, must be a perfect paradise for pigeons, since about half of the species known to science occur in that region only. The wonga-wonga and bronze-wing and great fruit-pigeons are, like the "bald-pates" of Jamaica, all favourite birds with sportsmen, and some of the birds are far more brightly coloured than ours. It is, however, noticeable that even the gayest Queensland species, with wings shot with every prismatic hue, are dull-looking birds seen from above, and the late Dr. A. R. Wallace regarded this as affording protection against keen-eyed hawks on the forage. His ingenious theory receives support from the well-known fact that in many of the islands, where pigeons are even more plentiful, but where also hawks are few, the former wear bright clothes on their back as well.

The woodpigeon has many names in rural England. That by which it is referred to in the foregoing notes is not, perhaps, the most satisfactory, since, with the possible exception [39] of the smaller stock-dove, which lays its eggs in rabbit burrows, and the rock-dove, which nests in the cliffs, all the members of the family need trees, if only to roost and nest in. A more descriptive name is that of ringdove, easily explained by the white collar, but the bird is also known as cushat, queest, or even culver. The last-named, however, which will be familiar to readers of Tennyson, probably alludes specifically to the rock-dove, as it undoubtedly gave its name to Culver Cliff, a prominent landmark in the Isle of Wight, where these birds have at all times been sparingly in evidence.

The ringdove occasionally rears a nestling in captivity, but it does not seem, at any time of life, to prove a very attractive pet. White found it strangely ferocious, and another writer describes it as listless and uninteresting. The only notable success on record is that scored by St. John, who set some of the eggs under a tame pigeon and secured one survivor that appears to have grown quite tame, but was, unfortunately, eaten by a hawk. At any rate, it did its kind good service by enlisting on their side the [40] pen of the most ardent apologist they have ever had. Indeed, St. John did not hesitate to rate the farmers soundly for persecuting the bird in wilful ignorance of its unpaid services in clearing their ground of noxious weeds. Yet, however true his eloquent plea may have been in respect of his native Lothian, there would be some difficulty in persuading South Country agriculturists of the woodpigeon's hidden virtues. To those, however, who do not sow that they may reap, the subject of these remarks has irresistible charm. There is doubtless monotony in its cooing, yet, heard in a still plantation of firs, with no other sound than perhaps the distant call of a shepherd or barking of a farm dog, it is a music singularly in harmony with the peaceful scene. The arrowy flight of these birds when they come in from the fields at sundown and fall like rushing waters on the treetops is an even more memorable sound. To the sportsman, above all, the woodpigeon shows itself a splendid bird of freedom, more cunning than any hand-reared game bird, swifter on the wing than any other purely wild bird, a [41] welcome addition to the bag because it is hard to shoot in the open, and because in life it was a sore trial to a class already harassed with their share of this life's troubles.

APRIL
BIRDS IN THE HIGH HALL GARDEN
   [45]

BIRDS IN THE HIGH HALL GARDEN

All March the rooks were busy in the swaying elms, but it is these softer evenings of April, when the first young leaves are beginning to frame the finished nests, and the boisterous winds of last month no longer drown the babble of the tree-top parliament at the still hour when farm labourers are homing from the fields, that the rooks peculiarly strike their own note in the country scene. There is no good reason to confuse these curious and interesting fowl with any other of the crow family. Collectively they may be recognised by their love of fellowship, for none are more sociable than they. Individually the rook is stamped unmistakably by the bald patch on the face, where the feathers have come away round the base of the beak. The most generally accepted explanation of this disfigurement is the rook's habit of thrusting its bill deep in the earth in search of its daily food. This, on the face of it, looks like a reasonable explanation, but it should be borne in mind that not only do [46] some individual rooks retain through life the feathers normally missing, but that several of the rook's cousins dip into Nature's larder in the same fashion without suffering any such loss. However, the featherless patch on the rook's cheeks suffices, whatever its cause, as a mark by which to recognise the bird living or dead.

Unlike its cousin the jackdaw, which commonly nests in the cliffs, the rook is not, perhaps, commonly associated with the immediate neighbourhood of the sea, but a colony close to my own home in Devonshire displays sufficiently interesting adaptation to estuarine conditions to be worth passing mention. Just in the same way that gulls make free of the wireworms on windswept ploughlands, so in early summer do the old rooks come sweeping down from the elms on the hill that overlooks my fishing ground and take their share of cockles and other muddy fare in the bank uncovered by the falling tide. Here, in company with gulls, turnstones, and other fowl of the foreshore, the rooks strut importantly up and down, digging their powerful bills deep in the ooze and occasionally [47] bullying weaker neighbours out of their hard-earned spoils. The rook is a villain, yet there is something irresistible in the effrontery with which one will hop sidelong on a gorging gull, which beats a hasty retreat before its sable rival, leaving some half-prized shellfish to be swallowed at sight or carried to the greedy little beaks in the treetops. While rooks are far more sociable than crows, the two are often seen in company, not always on the best of terms, but usually in a condition suggestive of armed neutrality. An occasional crow visits my estuary at low tide, but, though the bird would be a match for any single rook, I never saw any fighting between them. Possibly the crow feels its loneliness and realises that in case of trouble none of its brothers are there to see fair play. Yet carrion crows, like herons, are among the rook's most determined enemies, and cases of rookeries being destroyed by both birds are on record. On the other hand, though the heron is the far more powerful bird of the two, heronries have likewise been scattered, and their trees appropriated, by rooks, probably in overwhelming numbers. Of the two the [48] heron is, particularly in the vicinity of a preserved trout stream, the more costly neighbour. Indeed it is the only other bird which nests in colonies of such extent, but there is this marked difference between herons and rooks, that the former are sociable only in the colony. When away on its own business, the heron is among the most solitary of birds, having no doubt, like many other fishermen, learnt the advantage of its own company.

One of the most remarkable habits in the rook is that of visiting the old nests in mid-winter. Now and again, it is true, a case of actually nesting at that season has been noticed, but the fancy for sporting round the deserted nests is something quite different from this. I have watched the birds at the nests on short winter days year after year, but never yet saw any confirmation of the widely accepted view that their object is the putting in order of their battered homes for the next season. It seems a likely reason, but in that case the birds would surely be seen carrying twigs for the purpose, and I never saw them do so before January. What other attraction the empty nurseries can [49] have for them is a mystery, unless indeed they are sentimental enough to like revisiting old scenes and cawing over old memories.

The proximity of a rookery does not affect all people alike. Some who, ordinarily dwelling in cities, suffer from lack of bird neighbours, would regard the deliberate destruction of a rookery as an act of vandalism. A few, as a matter of fact, actually set about establishing such a colony where none previously existed, an ambition that may generally be accomplished without extreme difficulty. All that is needed is to transplant a nest or two of young rooks and lodge them in suitable trees. The parent birds usually follow, rear the broods, and forthwith found a settlement for future generations to return to. Even artificial nests, with suitable supplies of food, have succeeded, and it seems that the rook is nowhere a very difficult neighbour to attract and establish.

Why are rooks more sociable than ravens, and what do they gain from such communalism? These are favourite questions with persons informed with an intelligent passion for acquiring information, and the best [50] answer, without any thought of irreverence, is "God knows!" It is most certain that we, at any rate, do not. So far from explaining how it was that rooks came to build their nests in company, we cannot even guess how the majority of birds came to build nests at all, instead of remaining satisfied with the simpler plan of laying their eggs in the ground that is still good enough for the petrels, penguins, kingfishers, and many other kinds. Protection of the eggs from rain, frost, and natural enemies suggests itself as the object of the nest, but the last only would to some extent be furthered by the gregarious habit, and even so we have no clue as to why it should be any more necessary for rooks than for crows. To quote, as some writers do, the numerical superiority of rooks over ravens as evidence of the benefits of communal nesting is to ignore the long hostility of shepherds towards the latter birds on which centuries of persecution have told irreparably. Rooks, on the other hand, though also regarded in some parts of these islands as suspects, have never been harassed to the same extent; and if anything in the nature [51] of general warfare were to be inaugurated against them, the gregarious habit, so far from being a protection, would speedily and disastrously facilitate their extermination. Another curious habit noticed in these birds is that of flying on fine evenings to a considerable height and then swooping suddenly to earth, often on their backs. These antics, comparable to the drumming of snipe and roding of woodcock, are probably to be explained on the same basis of sexual emotion.

The so-called parliament of the rooks probably owes much of its detail to the florid imagination of enthusiasts, always ready to exaggerate the wonders of Nature; but it also seems to have some existence in fact, and privileged observers have actually described the trial and punishment of individuals that have broken the laws of the commune. I never saw this procedure among rooks, but once watched something very similar among the famous dogs of Constantinople, which no longer exist.

The most important problem however in connection with the rook is the precise extent to which the bird is the farmer's enemy or his friend. On the solution hangs the rook's fate [52] in an increasingly practical age, which may at any moment put sentiment on one side and decree for it the fate that is already overtaking its big cousin the raven. Scotch farmers have long turned their thumbs down and regarded rooks as food for the gun, but in South Britain the bird's apologists have hitherto been able to hold their own and avert catastrophe from their favourite. The evidence is conflicting. On the one hand, it seems undeniable that the rook eats grain and potato shoots. It also snaps young twigs off the trees and may, like the jay and magpie, destroy the eggs of game birds. On the other hand, particularly during the weeks when it is feeding its nestlings, it admittedly devours quantities of wireworms, leathergrubs, and weevils, as well as of couch grass and other noxious weeds, while some of its favourite dainties, such as thistles, walnuts, and acorns, will hardly be grudged at any time. It is not an easy matter to decide; and, if the rook is to be spared, economy must be tempered with sentiment, in which case the evidence will perhaps be found to justify a verdict of guilty, with a strong recommendation to mercy.

MAY
THE CUCKOO
[55]

THE CUCKOO

With the single exception of the nightingale, bird of lovers, no other has been more written of in prose or verse than the so-called "harbinger of spring." This is a foolish name for a visitor that does not reach our shores before, at any rate, the middle of April. Even *Whitaker* allows us to recognise the coming of spring nearly a month earlier; and for myself, impatient if only for the illusion of Nature's awakening, I date my spring from the ending of the shortest day. Once the days begin to lengthen, it is time to glance at the elms for the return of the rooks and to get out one's fishing-tackle again. Yet the cuckoo comes rarely before the third week of April, save in the fervent imagination of premature heralds, who, giving rein to a fancy winged by desire, or honestly deceived by some village cuckoo clock heard on their country rambles, solemnly write to the papers announcing the inevitable March cuckoo. They know better in the Channel Islands, for in the second week [56] of April, and not before, there are cuckoos in every bush—hundreds of exhausted travellers pausing for strength to complete the rest of their journey to Britain. Not on the return migration in August do the wanderers assemble in the islands, since, having but lately set out, they are not yet weary enough to need the rest. The only district of England in which I have heard of similar gatherings of cuckoos is East Anglia, where, about the time of their arrival, they regularly collect in the bushes and indulge in preliminary gambols before flying north and west.

Cuckoos, then, reach these islands about the third week of April, and they leave us again at the end of the summer, the old birds flying south in July, the younger generation following three or four weeks later. Goodness knows by what extraordinary instinct these young ones know the way. But the young cuckoo is a marvel altogether in the manner of its education, since, when one comes to think of it, it has no upbringing by its own parents and cannot even learn how to cry "Cuckoo!" by example or instruction. Its foster-parents speak another language, and [57] its own folk have ceased from singing by the time it is out of the nest. A good deal has been written about the way in which the note varies, chiefly in the direction of greater harshness and a more staccato and less sustained note, towards the end of the cuckoo's stay. According to the rustic rhyme, it changes its tune in June, which is probably poetic licence rather than the fruits of actual observation. It is, however, commonly agreed that the cuckoo is less often heard as the time of its departure draws near, and the easiest explanation of its silence, once the breeding season is ended, is that the note, being the love-call of a polygamous bird, is no longer needed.

In Australia the female cuckoo is handsomely barred with white, whereas the male is uniformly black; but with our bird it is exceedingly difficult to distinguish one sex from the other on the wing, and, were it not for occasional evidence of females having been shot when actually calling, we might still believe that it is the male only that makes this sound. The note is joyous only in the poet's fancy, just as he has also read sadness [58] into the "sobbing" of the nightingale. There is, indeed, when we consider its life, something fantastic in the hypothesis that the cuckoo can know no trouble in life, merely because it escapes the rigours of our winter. Eternal summer must be a delight, but the cuckoo has to work hard for the privilege, and it must at times be harried to the verge of desperation by the small birds that continually mob it in broad daylight. This behaviour on the part of its pertinacious little neighbours has been the occasion of much futile speculation; but the one certain result of such persecution is to make the cuckoo, along with its fellow-sufferer, the owls, preferably active in the sweet peace of the gloaming, when its puny tyrants are gone to roost. Much heated argument has raged round the real or supposed sentiment that inspires such demonstrations on the part of linnets, sparrows, chaffinches, and other determined hunters of the cuckoo. It seems impossible, when we observe the larger bird's unmistakable desire to win free of them, to attribute friendly feelings to its pursuers. Yet some writers have held the [59] curious belief that, with lingering memories of the days when, a year ago, they devoted themselves to the ugly foster-child, the little birds still regard the stranger with affection. If so, then they have an eccentric way of showing it, and the cuckoo, driven by the chattering little termagants from pillar to post, may well pray to be saved from its friends. On the other hand, even though convinced of their hostility, it is not easy to believe, as some folks tell us, that they mistake the cuckoo for a hawk. Even the human eye, though slower to take note of such differences, can distinguish between the two, and the cuckoo's note would still further undeceive them. The most satisfactory explanation of all perhaps is that the nest memories do in truth survive, not, however, investing the cuckoo with a halo of romance, but rather branding it as an object of suspicion, an interloper, to be driven out of the neighbourhood at all costs ere it has time to billet its offspring on the hardworking residents. All of which is, needless to say, the merest guesswork, since any attempt to interpret the simplest actions of birds is likely to lead us [60] into erroneous conclusions. Yet, of the two, it certainly seems more reasonable to regard the smaller birds as resenting the parasitic habit in the cuckoo than to admit that they can actually welcome the murder of their own offspring to make room in the nest for the ugly changeling foisted on them by this fly-by-night.

On the *lucus a non lucendo* principle, the cuckoo is chiefly interesting as a parent. The bare fact is that our British kind builds no nest of its own, but puts its eggs out to hatch, choosing for the purpose the nests of numerous small birds which it knows to be suitable. Further investigation of the habits of this not very secretive bird, shows that she first lays her egg on the ground and then carries it in her bill to a neighbouring nest. Whether she first chooses the nest and then lays the egg destined to be hatched in it, or whether she lays each egg when so moved and then hunts about for a home for it, has never been ascertained. The former method seems the more practical of the two. On the other hand, little nests of the right sort are so plentiful in May that, with her mother-instinct to guide [61] her, she could always find one at a few moments' notice. Some people, who are never so happy as when making the wonders of Nature seem still more wonderful than they really are, have declared that the cuckoo lays eggs to match those among which she deposits them, or that, at any rate, she chooses the nests of birds whose eggs approximately resemble her own. I should have liked to believe this, but am unfortunately debarred by the memory of about forty cuckoo's eggs that I took, seven-and-twenty summers ago, in the woods round Dartford Heath. The majority of these were found in hedgesparrows' nests, and the absolute dissimilarity between the great spotted egg of the cuckoo and the little blue egg of its so-called dupe would have impressed even a colour-blind animal. Occasionally, I believe, a blue cuckoo's egg has been found, but such a freak could hardly be the result of design. As a matter of fact, there is no need for any such elaborate deception. Up to the moment of hatching, the little foster-parents have in all probability no suspicion of the trick that has been played on them. Birds do not take [62] deliberate notice of the size or colour of their own eggs. Kearton somewhere relates how he once induced a blackbird to sit on the eggs of a thrush, and a lapwing on those of a redshank. So, too, farmyard hens will hatch the eggs of ducks or game birds and wild birds can even be persuaded to sit on eggs made of painted wood. Why then, since they are so careless of appearances, should the cuckoo go to all manner of trouble to match the eggs of hedgesparrow, robin or warbler? The bird would not notice the difference, and, even if she did, she would probably sit quite as close, if only for the sake of the other eggs of her own laying. Once the ugly nestling is hatched, there comes swift awakening. Yet there is no thought of reprisal or desertion. It looks rather as if the little foster-parents are hypnotised by the uncouth guest, for they see their own young ones elbowed out of the home and continue, with unflagging devotion, to minister to the insatiable appetite of the greedy little murderer. A bird so imbued as the parasitic cuckoo with the *Wanderlust* would make a very careless parent, and we must therefore perhaps revise our unflattering [63] estimate of its attitude and admit that it does the best it can by its offspring in putting them out to nurse. This habit, unique among British birds, is practised by many others elsewhere, and in particular by the American troupials, or cattle-starlings. One of these indeed goes even farther, since it entrusts its eggs to the care of a nest-building cousin. There are also American cuckoos that build their own nest and incubate their own eggs.

On the whole, our cuckoo is a friend to the farmer, for it destroys vast quantities of hairy caterpillars that no other bird, resident or migratory, would touch. On the other hand, no doubt, the numbers of other small useful birds must suffer, not alone because the cuckoo sucks their eggs, but also because, as has been shown, the rearing of every young cuckoo means the destruction of the legitimate occupants of the nest. So far however as the farmer is concerned, this is probably balanced by the reflection that a single young cuckoo is so rapacious as to need all the insect food available.

The cuckoo, like the woodcock, is supposed to have its forerunner. Just as the small [64] horned owl, which reaches our shores a little in advance of the latter, is popularly known as the "woodcock owl," so also the wryneck, which comes to us about the same time as the first of the cuckoos, goes by the name of "cuckoo-leader." It is never a very conspicuous bird, and appears to be rarer nowadays than formerly. Schoolboys know it best from its habit of hissing like a snake and giving them a rare fright when they cautiously insert a predatory hand in some hollow tree in search of a possible nest. It is in such situations that, along with titmice and some other birds, the wryneck rears its young; and it doubtless owes many an escape to this habit of hissing, accompanied by a vigorous twisting of its neck and the infliction of a sufficient peck, easily mistaken in a moment of panic for the bite of an angry adder. Thus does Nature protect her weaklings.

JUNE
VOICES OF THE NIGHT

[67]

VOICES OF THE NIGHT

The majority of nocturnal animals, more particularly those bent on spoliation, are strangely silent. True, frogs croak in the marshes, bats shrill overhead at so high a pitch that some folks cannot hear them, and owls hoot from their ruins in a fashion that some vote melodious and romantic, while others associate the sound rather with midnight crime and dislike it accordingly. The badger, on the other hand, with the otter and fox — all of them sad thieves from our point of view — have learnt, whatever their primeval habits, to go about their marauding in stealthy silence; and it is only in less settled regions that one hears the jackals barking, the hyænas howling, and the browsing deer whistling through the night watches.

There are, however, two of our native birds, or rather summer visitors, since they leave us in autumn, closely associated with these warm June nights, the stillness of which they break in very different fashion, and these are the nightingale and nightjar. Each is of considerable [68] interest in its own way. It is not to be denied that the churring note of the nightjar is, to ordinary ears, the reverse of attractive, and the bird is not much more pleasing to the eye than to the ear; while the nightingale, on the contrary, produces such sweet sounds as made Izaak Walton marvel what music God could provide for His saints in heaven when He gave such as this to sinners on earth. The suggestion was not wholly his own, since the father of angling borrowed it from a French writer; but he vastly improved on the original, and the passage will long live in the hearts of thousands who care not a jot for his instructions in respect of worms. At the same time, the nightjar, though the less attractive bird of the two, is fully as interesting as its comrade of the summer darkness, and there should be no difficulty in indicating the little that they have in common, as well as much wherein they differ, in both habits and appearance.

Both, then, are birds of sober attire. Indeed of the two, the nightjar, with its soft and delicately pencilled plumage and the conspicuous white spots, is perhaps the handsomer, [69] though, as it is seen only in the gloaming, its quiet beauty is but little appreciated. The unobtrusive dress of the nightingale, on the other hand, is familiar in districts in which the bird abounds, and is commonly quoted, by contrast with its unrivalled voice, as the converse of the gaudy colouring of raucous macaws and parrakeets. As has been said, both these birds are summer migrants, the nightingale arriving on our shores about the middle of April, the nightjar perhaps a fortnight later. Thenceforth, however, their programmes are wholly divergent, for, whereas the nightjars proceed to scatter over the length and breadth of Britain, penetrating even to Ireland in the west and as far north as the Hebrides, the nightingale stops far short of these extremes and leaves whole counties of England, as well as probably the whole of Scotland, and certainly the whole of Ireland, out of its calculations. It is however well known that its range is slowly but surely extending towards the west.

This curiously restricted distribution of the nightingale, indeed, within the limits of its summer home is among the most remarkable [70] of the many problems confronting the student of distribution, and successive ingenious but unconvincing attempts to explain its seeming eccentricity, or at any rate caprice, in the choice of its nesting range only make the confusion worse. Briefly, in spite of a number of doubtful and even suspicious reports of the bird's occurrence outside of these boundaries, it is generally agreed by the soundest observers that its travels do not extend much north of the city of York, or much west of a line drawn through Exeter and Birmingham. By way of complicating the argument, we know, on good authority, that the nightingale's range is equally peculiar elsewhere; and that, whereas it likewise shuns the departments in the extreme west of France, it occurs all over the Peninsula, a region extending considerably farther into the sunset than either Brittany or Cornwall, in both of which it is unknown. No satisfactory explanation of the little visitor's objection to Wild Wales or Cornwall has been found, and it may at once be stated that its capricious distribution cannot be accounted for by any known facts of soil, [71] climate, or vegetation, since the surroundings which it finds suitable in Kent and Sussex are equally to be found down in the West Country, but fail to attract their share of nightingales.

The song of the nightingale, in praise of which volumes have been written, is perhaps more beautiful than that of any other bird, though I have heard wonderful efforts from the mocking-bird in the

United States and from the bulbuls along the banks of the Jordan. The latter are sometimes, more especially in poetry, regarded as identical with the nightingale; and, indeed, some ornithologists hold the two to be closely related. What a gap there is between the sobbing cadences of the nightingale and the rasping note of the nightjar, which, with specific reference to a Colonial cousin of that bird Tasmanians ingeniously render as "more pork"! It seems almost ludicrous to include under the head of birdsong not only the music of the nightingale, but also the croak of the raven and the booming note of the ostrich. Yet these also are the love-songs of their kind, and the hen ostrich doubtless [72] finds more music in the thunderous note of her lord than in the faint melody of such song-birds as her native Africa provides. The nightingale sings to his mate while she is sitting on her olive-green eggs perching on a low branch of the tree, at foot of which the slender nest is hidden in the undergrowth. So much is known to every schoolboy who is too often guided by the sound on his errand of plunder; and why the song of this particular warbler should have been described by so many writers as one of sadness, seeing that it is associated with the most joyous days in the bird's year, passes comprehension. So obviously is its object to hearten the female in her long and patient vigil that as soon as the young are hatched the male's voice breaks like that of other choristers to a guttural croak. It is said, indeed—though so cruel an experiment would not appeal to many—that if the nest be destroyed just as the young are hatched the bird recovers all his sweetness of voice and sings anew while another home is built.

Although poetic licence has ascribed the song to the female, it is the male nightingale [73] only that sings, and for the purpose aforementioned. The note of the nightjar, on the other hand, is equally uttered by both sexes, and both also have the curious habit of repeatedly clapping the wings for several minutes together. They moreover share the business of incubation, taking day and night duty on the eggs, which, two in number, are laid on the bare ground without any pretence of a nest, and generally on open commons in the neighbourhood of patches of fern-brake. Like the owls, these birds sleep during the day and are active only when the sun goes down. It is this habit of seeking their insect food only in the gloaming which makes nightjars among the most difficult of birds to study from life, and all accounts of their feeding habits must therefore be received with caution, particularly that which compares the bristles on the mouth with baleen in whales, serving as a sort of strainer for the capture of minute flying prey. This is an interesting suggestion, and may even be sober fact; but its adoption would necessitate the bird flying open-mouthed among the oaks and other trees beneath which it finds [74] the yellow underwings and cockchafers on which it feeds, and I have more than once watched it hunting its victims with the beak closed. I noticed this particularly when camping in the backwoods of Eastern Canada where the bird goes by the name of nighthawk.

In all probability its food consists exclusively of insects, though exceptional cases have been noted in which the young birds had evidently been fed on seeds. The popular error which charges it with stealing the milk of ewes and goats, from which it derives the undeserved name of "goat-sucker," with its equivalent in several Continental languages, is another result of the imperfect light in which it is commonly observed. Needless to say, there is no truth whatever in the accusation, for the nightjar would find no more pleasure in drinking milk than we should in eating moths.

Here, then, are two night-voices of very different calibre. These are not our only birds that break the silence on moonlight nights in June. The common thrush often sings far into the night, and the sedge-warbler is a [75] persistent caroller that has often been mistaken for the nightingale. The difference in this respect between the two subjects of these remarks is that the nightjar is invariably silent all through the day, whereas the nightingale sings joyously at all hours. It is only because his splendid music is more marked in the comparative silence of the night, with little or no competition, that his daylight concert is often overlooked.

JULY
SWIFTS, SWALLOWS AND MARTINS
[79]

SWIFTS, SWALLOWS AND MARTINS

When the trout-fisherman sees the first martins and swallows dipping over the sward of the water-meadows and skimming the surface of the stream in hot pursuit of such harried water-insects as have escaped the jaws of greedy fish, he knows that summer is coming in. The signs of spring have been evident in the budding hedgerows for some weeks. The rooks are cawing in the elms, the cuckoo's note has been heard in the spinney for some time before these little visitors pass in jerky flight up and down the valley. Then, a little later, come the swifts — the black and screaming swifts — which, though learned folk may be right in sundering them utterly from their smaller travelling companions from the sunny south, will always in the popular fancy be associated with the rest. Colonies of swifts, swallows, and martins are a dominant feature of English village life during the warm months; and though there are fastidious folk who take not wholly culpable exception to their little visitors on [80] the score of cleanliness, most of us welcome them back each year, if only for the sake of the glad season of their stay. If, moreover, it is a question of choice between these untiring travellers resting in our eaves and the stay-at-home starling or sparrow, the choice will surely fall on the first every time.

The swift is the largest and most rapid in its flight, and its voice has a penetrating quality lacking in the notes of the rest. Swifts screaming in headlong flight about a belfry or up and down a country lane are the embodiment of that sheer joy of life which, in some cases with slender reason, we associate peculiarly with the bird-world. Probably, however, these summer migrants are as happy as most of their class. On the wing they can have few natural enemies, though one may now and again be struck down by a hawk; and they alight on the ground so rarely as to run little risk from cats or weasels, while the structure and position of their nests alike afford effectual protection for the eggs and young. Compared with that of the majority of small birds, therefore, their existence should be singularly happy and free [81] from care; and though that of the swift can scarcely, perhaps, when we remember its shrill voice, be described as one grand sweet song, it should not be chequered by many troubles. The greatest risk is no doubt that of being snapped up by some watchful pike if the bird skims too close to the surface of either still or running water, and I have even heard of their being seized in this way by hungry mahseer, those great barbel which gladden the heart of exiled anglers whose lot is cast on the banks of Himalayan rivers.

It is, however, the sparrows and starlings, rivals for the nesting sites, who show themselves the irreconcilable enemies of the returned prodigals. Terrific battles are continually enacted between them with varying fortunes, and the anecdotes of these frays would fill a volume. Jesse tells of a feud at Hampton Court, in the course of which the swallows, having only then completed their nest, were evicted by sparrows, who forthwith took possession and hatched out their eggs. Then came Nemesis, for the sparrows were compelled to go foraging for food with which to fill the greedy beaks, and during their [82] enforced absence the swallows returned in force, threw the nestlings out, and demolished the home. The sparrows sought other quarters, and the swallows triumphantly built a new nest on the ruins of the old. A German writer relates a case of revolting reprisal on the part of some swallows against a sparrow that appropriated their nest and refused to quit. After repeated failure to evict the intruder, the swallows, helped by other members of the colony, calmly plastered up the front door so effectually that the unfortunate sparrow was walled up alive and died of hunger. This refined mode of torture is not unknown in the history of mankind, but seems singularly unsuited to creatures so fragile.

The nests of these birds show, as a rule, little departure from the conventional plan, but they do adapt their architecture to circumstances, and I remember being much struck on one occasion by the absence of any dome or roof. It was in Asia Minor, on the seashore, that I came upon a cottage long deserted, its door hanging by one hinge, and all the glass gone from the windows. In the empty rooms numerous swallows were rearing [83] twittering broods in roofless nests. No doubt the birds realised that they had nothing to fear from rain, and were reluctant to waste time and labour in covering their homes with unnecessary roofs.

Most birds are careful in the education of their young, and indeed thorough training at an early stage must be essential in the case of creatures that are left to protect themselves and to find their own food when only a few weeks old. Fortunately they develop with a rapidity that puts man and other mammals to shame, and the helpless bald little swift lying agape in the nest will in another fortnight be able to fly across Europe. One of the most favoured observers of the early teaching given by the mother-swallow to her brood was an angler who told me how, one evening when he was fishing in some ponds at no great distance from London, a number of baby swallows alighted on his rod. He kept as still as possible, fearful of alarming his interesting visitors, but he must at last have moved, for, with one accord, they all fell off his rod together, skimmed over the surface of the water and disappeared in the direction [84] from which they had come a few moments earlier.

Swifts fly to an immense height these July evenings, mounting to such an altitude as eventually to disappear out of sight altogether. This curious habit, which is but imperfectly understood, has led to the belief that, instead of roosting in the nest or among the reeds like the swallows, the males, at any rate, spend the night flying about under the stars. This fantastic notion is not, however, likely to commend itself to those who pause to reflect on the incessant activity displayed by these birds the livelong day. So rarely indeed do they alight that country folk gravely deny them the possession of feet, and it is in the last degree improbable that a bird of such feverish alertness could dispense with its night's rest. No one who has watched swifts, swallows and martins on the wing can fail to be struck by the extraordinary judgment with which these untiring birds seem to shave the arches of bridges, gateposts, and other obstacles in the way of their flight by so narrow a margin as continually to give the impression of catastrophe imminent and [85] inevitable. Their escapes from collision are marvellous; but the birds are not infallible, as is shown by the untoward fate of a swallow in Sussex. In an old garden in that county there had for many years been an open doorway with no door, and through the open space the swallows had been wont, year after year, to fly to and fro on their hunting trips. Then came a fateful winter during which a new owner took it into his head to put up a fresh gate and to keep it locked, and, as ill luck would have it, he painted it blue, which, in the season of fine weather, probably heightened the illusion. Back came the happy swallows to their old playground, and one of the pioneers flew headlong at the closed gate and fell stunned and dying on the ground, a minor tragedy that may possibly come as a surprise to those who regard the instincts of wild birds as unerring.

That the young swallows leave our shores before their elders — late in August or early in September — is an established fact, and the instinct which guides them aright over land and sea, without assistance from those more experienced, is nothing short of amazing. [86] The swifts, last to come, are also first to go, spending less time in the land of their birth than either swallows or martins. The fact that an occasional swallow has been seen in this country during the winter months finds expression in the adage that "one swallow does not make a summer," and it was no doubt this occasional apparition that in a less enlightened age seemed to warrant the extraordinary belief, which still ekes out a precarious existence in misinformed circles, that these birds, instead of wintering abroad, retire in a torpid condition to the bottom of lakes and ponds. It cannot be denied that these waters have occasionally, when dredged or drained, yielded a stray skeleton of a swallow, but it should be evident to the most homely intelligence that such débris merely indicates careless individuals that, in passing over the water, got their plumage waterlogged and were then drowned. It seems strange that Gilbert White, so accurate an observer of birds, should actually have toyed with this curious belief, though he leant rather to the more reasonable version of occasional hybernation in caves or other sheltered [87] hiding-places. The rustic mind, however, preferred, and in some unsophisticated districts still prefers, the ancient belief in diving swallows, and no weight of evidence, however carefully presented, would shake it in its creed. Fortunately this eccentric view of the swallow's habits brings no harm to the bird itself, and may thus be tolerated as an innocuous indulgence on the part of those who prefer this fiction to the even stranger truth.

AUGUST
THE SEAGULL
   [91]

THE SEAGULL

So glorious is the flight of the seagull that it tempts us to fling aside the dry-as-dust theories of mechanism of flexed wings, coefficient of air resistance, and all the abracadabra of the mathematical biologist, and just to give thanks for a sight so inspiring as that of gulls ringing high in the eye of the wind over hissing combers that break on sloping beaches or around jagged rocks. These birds are one with the sea, knowing no fear of that protean monster which, since earth's beginning, has always, with its unfathomable mystery, its insatiable cruelty, its tremendous strength, been a source of terror to the land animals that dwell in sight of it. Yet the gulls sit on the curling rollers as much at their ease as swimmers in a pond, and give an impression of unconscious courage very remarkable in creatures that seem so frail. Hunger may drive them inland, or instincts equally irresistible at the breeding season, but never the worst gale that lashes the sea to fury, for they dread it in its hour [92] of rage as little as on still summer nights when, in their hundreds, they fly off the land to roost on the water outside the headlands.

It is curious that there should be no mention of them in the sacred writings. We read of quails coming in from the sea, likewise of "four great beasts," but of seafowl never a word, though one sees them in abundance on the coast near Jaffa, and the Hebrew writers might have been expected to weave them into the rich fabrics of their poetic imagery as they did the pelican, the eagle and other birds less familiar. Although seagulls have of late years been increasingly in evidence beside the bridges of London, they are still, to the majority of folk living far inland, symbolical of the August holiday at the coast, and their splendid flight and raucous cries are among the most enduring memories of that yearly escape from the smoke of cities.

The voice of gulls can with difficulty be regarded as musical, yet those of us who live the year round by the sea find their plaintive mewing as nicely tuned to that wild environment as the amorous gurgling of nightingales to moonlit woods in May. Their voice [93] may have no great range, but at any rate it is not lacking in variety, suggesting to the playful imagination laughter, tears, and other human moods to which they are in all probability strangers. The curious similarity between the note of a seagull and the whining of a cat bereft of her kittens is very striking, and was on one occasion the cause of my being taken in by one of these birds in a deep and beautiful backwater of the Sea of Marmora, beside which I spent one pleasant summer. In this particular gulf, at the head of which stands the ancient town of Ismidt, gulls, though plentiful in the open sea, are rarely in evidence, being replaced by herons and pelicans. I had not therefore set eyes on a seagull for many weeks, when early one morning I heard, from the farther side of a wooded headland, a new note suggestive of a wild cat or possibly a lynx. My Greek servant tried in his patois to explain the unseen owner of the mysterious voice, but it was only when a small gull suddenly came paddling round the corner that I realised my mistake.

In addition to being at home on the seashore, [94] and particularly in estuaries and where the coast is rocky, gulls are a familiar sight in the wake of steamers at the beginning and ending of the voyage, as well as following the plough and nesting in the vicinity of inland meres and marshes. The black-headed kind is peculiarly given to bringing up its family far from the sea, just as the salmon ascends our rivers for the same purpose. It is not perhaps a very loving parent, seeing that the mortality among young gulls, many of which show signs of rough treatment by their elders, is unusually great. On most lakes rich in fish these birds have long established themselves, and they were, I remember, as familiar at Geneva and Neuchâtel as along the shores of Lake Tahoe in the Californian Sierras, itself two hundred miles from the Pacific and more than a mile above sea-level. Gulls also follow the plough in hordes, not always to the complete satisfaction of the farmer, who is, not unreasonably, sceptical when told that they seek wireworms only and have no taste for grain. Unfortunately the ordinary scarecrow has no terror for them, and I recollect, in the neighbourhood [95] of Maryport, seeing an immense number of gulls turning up the soil in close proximity to several crows that, dangling from gibbets, effectually kept all black marauders away.

Young gulls are, to the careless eye, apt to look larger than their parents, an illusion possibly due to the optical effect of their dappled plumage, and few people unfamiliar with these birds in their succeeding moults readily believe that the dark birds are younger than the white. Down in little Cornish harbours I have sometimes watched these young birds turned to good account by their lazy elders, who call them to the feast whenever the ebbing tide uncovers a heap of dead pilchards lying in three or four feet of water, and then pounce on them the moment they come to the surface with their booty. The fact is that gulls are not expert divers. The cormorant and puffin and guillemot can vanish at the flash of a gun, reappearing far from where they were last seen, and can pursue and catch some of the swiftest fishes under water. Some gulls, however, are able to plunge farther below the surface than others, and the little kittiwake is perhaps the [96] most expert diver of them all, though in no sense at home under water like the shag. I have often, when at anchor ten or fifteen miles from the land, and attended by the usual convoy of seabirds that invariably gather round fishing-boats, amused myself by throwing scraps of fish to them and watching the gulls do their best to plunge below the surface when some coveted morsel was going down into the depths, and now and again a little Roman-nose puffin would dive headlong and snatch the prize from under the gulls' eyes. Most of the birds were fearless enough; only an occasional "saddleback"—the greater black-backed gull of the text-books—knowing the hand of man to be against it for its raids on game and poultry, would keep at a respectful distance.

Considered economically, the smaller gulls at any rate have more friends than enemies, and they owe most of the latter not so much to their appetites, which set more store by offal and carrion than by anything of greater value, as to their exceedingly dirty habits. These unclean fowl are in fact anything but welcome in harbours given over in summer [97] to smart yachting craft; and I remember how at Avalon, the port of Santa Catalina Island (Cal.), various devices were employed to prevent them alighting. Boats at their moorings were festooned with strips of bunting, which apparently had the requisite effect, and the railings of the club were protected by a formidable armour of nails. On the credit side of their account with ourselves, seagulls are admittedly assiduous scavengers, and their services in keeping little tidal harbours clear of decaying fish which, if left to accumulate, would speedily breed a pestilence, cannot well be overrated. The fishermen, though they rarely molest them, do not always refer to the birds with the gratitude that might be expected, yet they are still further in their debt, being often apprised by their movement of the whereabouts of mackerel and pilchard shoals, and, in thick weather, getting many a friendly warning of the whereabouts of outlying rocks from the hoarse cries of the gulls that have their haunts on these menaces to inshore navigation.

Seagulls are not commonly made pets of, the nearest approach to such adoption being [98] an occasional pinioned individual enjoying qualified liberty in a backyard. Their want of popularity is easily understood, since they lack the music of the canary and the mimicry of parrots. That they are, however, capable of appreciating kindness has been demonstrated by many anecdotes. The Rev. H. A. Macpherson used to tell a story of how a young gull, found with a broken wing by the children of some Milovaig crofters, was nursed back to health by them until it eventually flew away. Not long after it had gone, one of the children was lost on the hillside, and the gull, flying overhead, recognised one of its old playmates and hovered so as to attract the attention of the child. Then, on being called, the bird settled and roosted on the ground beside him. An even more remarkable story is told of a gull taken from the nest, on the coast of county Cork, and brought up by hand until, in the following spring, it flew away in the company of some others of its kind that passed over the garden in which it had its liberty. The bird's owner reasonably concluded that he had seen the last of his protégée, and great was his astonishment [99] when, in the first October gale, not only did the visitor return, tapping at the dining-room window for admission, as it had always done, but actually brought with it a young gull, and the two paid him a visit every autumn for a number of years.

On either side of the gulls, and closely associated with them in habits and in structure, is a group of birds equally characteristic of the open coast, the skuas and terns. The skuas, darker and more courageous birds, are familiar to those who spend their August holiday sea-fishing near the Land's End, where, particularly on days when the east wind brings the gannets and porpoises close inshore, the great skua may be seen at its favourite game of swooping on the gulls and making them disgorge or drop their launce or pilchard, which the bird usually retrieves before it reaches the water. This act of piracy has earned for the skua its West Country sobriquet of "Jack Harry," and against so fierce an onslaught even the largest gull, though actually of heavier build than its tyrant, has no chance and seldom indeed seems to offer the feeblest resistance. These skuas rob their [100] neighbours in every latitude; and even in the Antarctic one kind, closely related to our own, makes havoc among the penguins, an episode described by the late Dr. Wilson, one of the heroes of the ill-fated Scott expedition.

Far more pleasing to the eye are the graceful little terns, or "sea-swallows," fairylike creatures with red legs and bill, long pointed wings and deeply forked tail, which skim the surface of the sea or hawk over the shallows of trout streams in search of dragonflies or small fish. It is not a very rare experience for the trout-fisherman to hook a swallow which may happen to dash by at the moment of casting; but a much more unusual occurrence was that of a tern, on a well-known pool of the Spey, actually mistaking a salmon-fly for a small fish and swooping on it, only to get firmly hooked by the bill. Fortunately for the too venturesome tern the fisherman was a lover of birds, and he managed with some difficulty to reel it in gently, after which it was released none the worse for its mistake.

SEPTEMBER
BIRDS IN THE CORN

[103]

BIRDS IN THE CORN

More than one of our summer visitors, like the nightingale and cuckoo, are less often seen than heard, but certainly the most secretive hider of them all is the landrail. This harsh-voiced bird reaches our shores in May, and it was on the last of that month that I lately heard its rasping note in a quiet park not a mile out of a busy market town on the Welsh border, and forgave its monotone because, more emphatically than even the cuckoo's dissyllable, it announced that, at last, "summer was icumen in." This feeble-looking but indomitable traveller is closely associated during its visit with the resident partridge. They nest in the same situations, hiding in the fields of grass and standing corn, and eventually being flushed in company by September guns walking abreast through the cloverbud. Sport is not the theme of these notes, and it will therefore suffice to remark in passing on the curious manner in which even good shots, accustomed to bring down partridges with some approach to [104] certainty, contrive to miss these lazy, flapping fowl when walking them up. Dispassionately considered, the landrail should be a bird that a man could scarcely miss on the first occasion of his handling a gun; in cold fact, it often survives two barrels apparently untouched. This immunity it owes in all probability to its slow and heavy flight, since those whose eyes are accustomed to the rapid movement of partridges are apt to misjudge the allowance necessary for such a laggard and to fire in front of it. It is difficult to realise that, whereas the strong-winged partridge is a stay-at-home, the deliberate landrail has come to us from Africa and will, if spared by the guns, return there.

Perhaps the most curious and interesting habit recorded of the landrail is that of feigning death when suddenly discovered, a method of self-defence which it shares with opossums, spiders, and in fact other animals of almost every class. It will, if suddenly surprised by a dog, lie perfectly still and betray no sign of life. There is, however, at least one authentic case of a landrail actually dying of fright when suddenly seized, and it [105] is a disputed point whether the so-called pretence of death should not rather be regarded as a state of trance. Strict regard for the truth compels the admission that on the only occasion on which I remember taking hold of a live corncrake the bird, so far from pretending to be dead, pecked my wrist heartily.

Just as the countryfolk regard the wryneck as leader of the wandering cuckoos, and the short-eared owl as forerunner of the woodcocks, so the ancients held that the landrail performed the same service of pioneer to the quail on its long journeys over land and sea. Save in exceptional years, England is not visited by quail in sufficient numbers to lend interest to this aspect of a bird attractive on other grounds, but the coincidence of their arrival with us is well established.

The voice of the corncrake, easily distinguished from that of any other bird of our fields, may be approximately reproduced by using a blunt saw against the grain on hard wood. So loud is it at times that I have heard it from the open window of an express train, the noise of which drowned all other birdsong, [106] and it seems remarkable that such a volume of sound should come from a throat so slender. Yet the rasping note is welcome during the early days of its arrival, since, just as the cuckoo gave earlier message of spring, so the corncrake, in sadder vein, heralds the ripeness of our briefer summer.

The East Anglian name "dakker-hen" comes from an old word descriptive of the bird's halting flight; and indeed to see a landrail drop, as already mentioned, after flying a few yards, makes one incredulous when tracing its long voyages on the map. In the first place, however, it should be remembered that the bird does not drop back in the grass because it is tired, but solely because it knows the way to safety by running out of sight. In the second, the apparent weakness of its wings is not real. Quails have little round wings that look ill adapted to long journeys. I have been struck by this times and again when shooting quail in Egypt and Morocco, yet of the quail's fitness for travel there has never, since Bible days, been any question.

The landrail is an excellent table bird. Personally I prefer it to the partridge, but [107] this is perhaps praising it too highly. Legally of course it is "game," as a game licence must be held by anyone who shoots it; and, though protected in this country only under the Wild Birds Act, Irish law extends this by a month, so that it may not be shot in that country after the last day of January. Like most migratory birds, its numbers vary locally in different seasons, and its scarcity in Hampshire, to which White makes reference, has by no means been maintained of recent years, as large bags have been recorded in every part of that county.

The common partridge is—at any rate for the naturalist—a less interesting subject than its red-legged cousin, which seems to have been first introduced from France (or possibly from the island of Guernsey, where it no longer exists) in the reign of Charles II. That this early experiment was not, however, attended by far-reaching results seems probable, since early in the reign of George III we find the Marquis of Hertford and other well-known sporting landowners making fresh attempts, the stock of "Frenchmen" being renewed from time to time during the next [108] fifty years, chiefly on the east side of England, where they have always been more in evidence than farther west. In Devon and Cornwall, indeed, the bird is very rare, and in Ireland almost unknown.

Its red legs stand it in good stead, for it can run like a hare, and in this way it often baffles the guns. It is not, however, so much its reluctance to rise that has brought it into disrepute with keepers as its alleged habit of ousting the native bird, in much the same way as the "Hanover" rat has superseded the black aboriginal, although far from the "Frenchman" driving the English partridge off the soil, there appears to be even no truth in the supposed hostility between the two, since they do not commonly affect the same type of country; and even when they meet they nest in close proximity and in comparative harmony. Nevertheless the males, even of the same species, are apt to be pugnacious in the breeding season.

Both the partridge and landrail run serious risk from scythe and plough while sitting on the nest. Landrails have before now been decapitated by the swing of the scythe, and [109] a case is on record in which a sitting partridge, seeing that the plough was coming dangerously near her nest, actually removed the whole clutch of eggs, numbering over a score, to the shelter of a neighbouring hedge. This was accomplished, probably with the help of the male, during the short time it took the plough to get to the end of the field and back, and is a remarkable illustration of devotion and ingenuity. Not for nothing indeed is the partridge a game bird, for it has been seen to attack cats, and even foxes, in defence of the covey; and I have seen, in the MS. notes of the second Earl of Malmesbury, preserved in the library at Heron Court, mention of one that drove off a carrion crow that menaced the family. Both partridge and landrail sit very close, particularly when the time of hatching is near, and Charles St. John saw a partridge, which his dog, having taken off the nest, was forced to drop, none the worse for her adventure, go straight back to her duties; though, as he adds, if it had not been that she knew that the eggs were already chipping she would in all probability have deserted her post for good and all. [110]

Whether or not France is to be regarded as the original home of the "red leg," the fact remains that in that country it is becoming scarcer every year, its numbers being maintained only in Brittany, Calvados, Orne, and Sarthe. Its distribution in Italy is equally capricious, for it is virtually restricted to the rocky slopes of the Apennines, the Volterrano Hills in Tuscany, and the coast ranges of Elba. It seems therefore that in Continental countries, as well as with us, the bird extends its range reluctantly. Game-preservers seem, however, to agree that partridges and pheasants are, beyond a certain point, incompatible as, with a limited supply of natural food, the smaller bird goes to the wall. Like most birds, partridges grow bold when pressed by cold and hunger, and I recollect hearing of a large covey being encountered ten or twelve years ago in an open space in the heart of the city of Frankfort.

OCTOBER
THE MOPING OWL

[113]

THE MOPING OWL

Music, vocal or otherwise, is always a matter of taste, and individual appreciation of birdsong varies like the rest. One man finds the cuckoo's cry intolerably wearisome. Another sees no romance in the gargling of doves, while comparatively few care for the piercing scream of the starling or the rasping note of the corncrake. Yet few birds perform to a more hostile audience than the owl. I say advisedly "the owl," since the vast majority of people make no distinction whatever between our three resident kinds of owl, not to mention at least half a dozen more visitors. Some excuse for such carelessness might perhaps be found in the similar flight and habits of different owls, but it might have been thought that greater measure of individual recognition on their own merits would have been conceded to birds that range in size from the dimensions of a sparrow to those of a duck. But no; an owl is just an owl. Why the soft and haunting cry of [114] these birds should not merely displease, but actually alarm, so many people unaccustomed to such sounds of the gloaming and darkness it would be difficult to say; but the voice of owls may possibly owe some of its disturbing effect to contrast with their silent flight, which, thanks to their fluffy plumage, with its broad quills and long barbs, prevents their making much more noise than ghosts when hunting rats and mice in moonlit fields. Only one other English bird has so quiet a flight, and that is the nightjar, another creature of the darkness, which, though no cousin to these nocturnal birds of prey, is known in some parts of the country as the "fern-owl." Visitors unprepared for the eerie woodland music of these autumn nights shudder when they hear the cry of the owl, as if it suggested midnight crime. For myself I have more agreeable associations, since I never hear one of these birds without recalling a gallant fight I once had with a big Tweed salmon in the weak light of a young moon, while three owls hooted amid the ghostly ruins of Norham Castle. Yet, even apart from this wholly agreeable memory, I find nothing unpleasant [115] in their music, and can readily conceive that the moping owl may sing to his mate as passionately as Philomel.

Not only is there the popular lack of distinction between one owl and another already referred to, but scientific ornithologists have displayed similar want of finality in classifying these birds. There are (as in seals) eared and earless owls, though the so-called "ears" in the birds are not actually ears at all, but tufts of feathers that give rather the impression of horns. There are bare-legged owls and owls with feather stockings. There are owls that fly by day and owls that fly by night, though this is a less satisfactory distinction than that between the diurnal butterflies and nocturnal moths. Any reliable classification of owls must, in short, rest on certain structural bony differences of interest only to the student of anatomy. Nearly all these birds are able to turn the outer toe completely round, and most of them, also, have very keen hearing, which must be an invaluable aid when hunting small animals in the dark.

Did the ancients actually regard the owl [116] as a wise bird, or was the fashion of depicting it in the following of Minerva merely dictated by the presence of these birds on the Akropolis? It seems hardly conceivable that they could so have blundered as to call the owls that we know clever birds; and the alternative assumption that owlish intellect can have appreciably changed in the interval is even less acceptable. It is probable that too much significance need not be attached to such association between the Greek goddess of wisdom and her attendant owls, for Hindu symbolism represented Ganesa, god of wisdom, with the head of an elephant, yet that animal, which the natives of India know better than the men of any other race, has never figured in their folklore as a type noted for its cunning. About the owl as we know it to-day, with its spectacled face and blinking eyes, there is nothing strikingly intelligent, and schoolboy slang, in which the word does duty as synonymous with foolishness, discovers a more accurate appreciation of these birds.

Seen at its worst, when surprised in the glare of daylight and mobbed by a furious [117] rabble of little birds, an owl looks a helpless fool indeed, though this is not the proper moment to judge of the bird's possibilities under happier circumstances. Why these small fowl should bully it at all is one of those woodland problems that no one has yet solved. The first, and obvious, explanation is that they know it for their enemy, and it may be indeed that owls commit depredations on the nests of wild birds of which we, who academically regard their food as consisting of rats, bats and mice — or, in the case of larger species, of young game and leverets — have no inkling. If however such is the case, it is strange that the habit should have been overlooked by those who have paid close attention to this curious and interesting group. Bird-catchers, at any rate, without troubling to inquire into the reason, turn the instinct to profitable account, and in some parts of the country a stuffed owl is an important item of their stock-in-trade.

The majority of owls that either reside in or visit these islands are benefactors of the farmer, and should be spared. The larger eagle-owl, and snowy owl eat more expensive [118] food, though, seeing that they come to us—at any rate in the south country—only in winter, and even then irregularly, they can do no damage to young game birds, and are probably incapable of capturing old. The worst offender among the residents is the tawny owl, to which I find the following reference in the famous Malmesbury MSS.: "Common here ... a great destroyer of young game and leverets ... they sit in ivy bushes during the day, and I have known one remain, altho' its mate was killed, in the same tree, in such a state of torpor did it appear to be...." The screech owl is a harmless bird and a terror to mice, and any doubt as to its claim on the farmer's hospitality would at once be removed by cursory examination of the undigested pellets which, in common with hawks, these birds cast up after their meals.

On the other hand, there is sometimes good reason for modifying any plea for kindness to owls. Handsome is as handsome does, and many of these birds are, during the nesting season, not only savage in defence of their young, but actually so aggressive [119] as to make unprovoked attack on all and sundry who unwittingly approach closer to the tree than these devoted householders think desirable. Accounts of this troublesome mood in nesting owls come from several parts of the country, and notably from Wales. In one case on record a pair of barn owls had their home in a tree overlooking Milford Haven, and the vicinity of the nest soon became dangerous. The male owl tore a boy's ear, knocked a man down, and attacked numerous human beings and dogs that made use of a path leading past the tree; and these episodes were in fact of daily occurrence until some one shot the bird. Another pair of barn owls nested in a wood on the shore of Menai Strait, and in this case the young birds managed to fall out of the nest, and lay on the ground in full view of a public right of way. Why the old birds did not put their offspring back in the nest no one knew. Possibly they realised that the talons, which so efficiently gripped rats, might not prove gentle enough for the transport of owlets. At any rate, whatever their reason, they left the young [120] birds on the ground, feeding them in that position, and flew at everyone who passed that way, clawing face and ears, and eventually establishing a reign of terror. Another owl behaved in somewhat similar fashion in a spinney close to Axmouth, South Devon, punishing a coastguard so severely that the man took to his heels. Such determined tactics in defence of the young are the more singular when we remember that owls are, in normal circumstances, shy and retiring birds. Yet they occasionally seem to be possessed by more sociable instincts, in proof of which one of the long-eared kind has been seen feeding in the company of tame hawks; a pair of owls once nested in a dovecote close to a keeper's lodge in the Highlands; and wild owls have been known to pay nightly visits to a cage in the Botanic Gardens at Launceston (Tasmania), in order to bring food to their captive friends.

Even apart from these rigorous measures of defence, the nesting habits of owls are not without interest. The majority lay their eggs in either hollow trees or ruins, and it is worth remark that these nocturnal [121] birds bring up their young in darkness, whereas the hawks—birds of daylight—rear theirs in open nests, high up in trees or on rocky ledges, in the full glare of the sun. One owl indeed habitually burrows in the prairies and pampas, in the curious company of marmots and rattlesnakes, and this burrowing habit is also, in some parts of the United States, adopted by the common barn owl. Owls generally brood from the laying of the first egg, with the obvious result that young birds in various stages of plumage are found together in the nest. It has been suggested that the body of the first to leave the egg helps to keep the unhatched eggs warm while the parents are away foraging, else its presence would be a serious handicap. The first little owl to hatch out is usually ready to leave the nest soon after the arrival of the last, though these chicks come into the world more helpless even than the majority of birds.

NOVEMBER
WATERFOWL

[125]

WATERFOWL

Had these notes been written from the standpoint of sport, the three familiar groups of birds, which together make up this worldwide aquatic family, might better have borne their alternative title "wildfowl" with its covert sneer at the hand-reared pheasant and artificially encouraged partridge that, between them, furnish so much comfortable sport to those with no fancy for the arduous business of the mudflats. It is true that, of late years, the mallard has, in experienced hands, made a welcome addition to the bag in covert shooting, as those will remember who have shot the Lockwood Beat on the last day of the shoot at Nuneham; and there is historic evidence of "wild" duck having been reared for purposes of sport with hawks in the reign of Charles I. Yet such armchair shooting of wildfowl was ignored by Colonel Hawker and the second Earl of Malmesbury, both of whom, gunning in the creeks and estuaries of the south coast, made immense bags of ducks and geese, working hard for every bird and displaying Spartan indifference to the rigours of wintry weather. To hardy sportsmen of their type, wildfowl offer red-letter days with punt or shoulder guns, not to be dreamt of under the ægis of the gamekeeper.

In this country, at any rate, we associate the V-shaped companies of wigeon and gaggles of geese with an ice-bound landscape, though in exceptional years, even where they no longer stay to breed, these night-flying northerners linger to the coming of spring, and Hawker noticed the curious apparition of grey geese and swallows in company on the first day of April, 1839. This wedge formation of flight over land and sea is not only peculiar to these waterfowl, but is not apparently adopted by any other long distance migrants. No satisfactory explanation of their preference for flying in this order has been found, but it is thought to lessen the air resistance, which must be a consideration for these short-pinioned fowl that weigh heavy in proportion to their displacement and at the same time lack the tremendous spread of wing that enables the wandering albatross to soar for days together over the illimitable ocean. With one noticeable exception, these waterfowl exhibit a more extraordinary range of size and weight than any other family of birds, from the whooper swan, five feet long and twenty-five pounds on the scales, down to the little teal, with an overall measurement of only fourteen inches and a weight that does not exceed as many ounces. The only other family of birds running to such extremes is that of the birds of prey, which include at once the stately condor of the Andes with its wing-spread of fifteen feet, and the miniature red-legged falconet of India and adjoining countries, in which the same measurement would scarcely reach as many inches.

Since even game birds are derisively referred to as "tame" only by those ignorant of the facts, the birds now under notice differ in this respect from all those previously dealt with; and they are geographically apart, again, from our other domesticated animals, since they are not, like the barndoor fowl and most of the rest, of Asiatic origin, but must often, in the grey of a winter morning, [128] be conscious of their near relations flying at liberty across the sky. The geese and ducks have been remarkably transformed by the process of domestication, and a comparison between those of the farmyard and their kindred in the marshes should illustrate not only the relative value of most virtues, but also the all-importance of Aristotle's how, when and where. Strictly speaking, no doubt, the tame birds have degenerated, both mentally and physically, as surely as the tame ass. They have lost the acute perceptions and swift flight of their wild relations. Economically, on the other hand, they are immeasurably improved, since the farmer, indifferent to the more inspiring personality of the grey goose and the mallard, merely wants his poultry to be greedy and stupid, fattening themselves incessantly for Leadenhall and easily captured when required.

Between swans, geese and ducks there is little anatomical difference, save in the matter of size. The swans are the giants of the race, and the swans of three continents are white. It was left for Australia, land of topsy-turveydom, to produce a black swan [129] (I spare the reader the obvious classical tag), and this remarkable bird, first observed by Europeans in the early days of 1697, was quickly brought to Europe and figures in the earliest list of animals shown in the London Zoological Gardens. All these birds have a curious trick of hissing when angry, and this habit, perhaps because it is usually accompanied by a deliberate stretching of the neck to its full length, is seriously regarded by some as conscious mimicry of snakes, a proposition that must be left to individual taste, but that strikes me as somewhat far-fetched. At any rate, it gives to these birds a formidable air, and, though the current belief in its power of breaking a man's arm with a blow from its wing is probably unwarranted, an angry swan, disturbed on its nest, is an awesome apparition of which I have twice taken hurried leave. On the first occasion, I had nothing but a valuable camera with me, and it was, in fact, after a futile attempt to photograph the bird on the nest that I was moved to seek the boat and push off from the little island in the Upper Thames on which it had its home. The other [130] encounter was on a Devonshire trout stream, and my only weapon was a fragile trout rod. The certainty that discretion is, under these circumstances, the better part of valour is emphasised by the knowledge that any violence to the bird would probably lead to a prosecution. Even the smaller geese can inspire fear when they dash hissing at intruders; hence, no doubt, the nursemaid's favourite reproach of children too frightened to "say bo to a goose," an expression made classical by Swift.

The majority of these waterfowl are insectivorous in the nursery stage and vegetarian when full grown. Fish forms an inappreciable portion of their food, with the two notorious exceptions of the goosander and merganser, though anglers are much exercised over the damage, real or alleged, done by these birds to their favourite roach and dace in the Thames. These swans belong for the most part to either the Crown or the Dyers' and Vintners' Companies, and the practice of "uppings," which consists in marking the beaks of adult birds and pinioning the cygnets, is still, though shorn of some [131] of its former ceremonial, observed some time during the month of June.

Swans, like both of the other groups, are distinguished by a separate name for either sex: pen and cob for the swan, gander and goose, drake and duck, and the figurative use of some of these terms in such popular sayings as "making ducks and drakes of money," "sauce for the goose," etc., is too familiar to call for more than passing mention.

Nearly all these waterfowl, though seen on dry land to much the same disadvantage as fish out of water, are exceedingly graceful in either air or water, though not all ducks are as capable of diving as the name would imply. The proverbial futility of a wild goose chase recognises the pace of these birds on the wing, which, though, in common with that of some other birds, popularly exaggerated, is considerably faster than, owing to their short wings and heavy build, might appear to the careless observer.

Ducks have a curious habit of adding down to the nest after the eggs are laid and before incubation, and this provision of warm packing is turned to account in Iceland and other [132] breeding places of the eider duck, commercially the most valuable of all ducks. The nest is robbed of this down once before the eggs hatch out, with the result that the female plucks another store from her own breast, supplemented if necessary from the body of the drake. The sitting bird is then left in peace till the nest has fulfilled its purpose, when the remaining down is likewise removed. This down, which combines warmth and lightness, gives a high market value to the eider, which, throughout Scandinavian countries is strictly protected by law and even more effectually by public opinion.

The majority of ornamental ducks interbreed freely in captivity. Those who, apparently on reliable evidence, distinguish between the polygamous habit in tame ducks and the constancy of the mallard and other wild kinds to a single mate have hastily assumed that such hybrids are unknown in the natural state. This, however, is incorrect, as there have been authentic cases of crosses between mallard and teal, pochard and scaup and other species, such hybrids having at [133] different times been erroneously accepted as distinct species and named accordingly.

The wild duck's nest is usually placed on the ground in some sheltered spot close to still or running water, and the ducklings swim like corks, soon learning the proper use of their flat little bills in gobbling up floating insects and other waterlogged food. Occasionally ducks nest in trees and they have been known to take possession of a deserted rook's nest. There has been some discussion as to whether, in this case, the mother conveys her ducklings to the water in her bill, but this has not actually been witnessed. In cases where, as is often observed, the nest overhangs the water, it has been suggested that the young birds may simply be pushed over the edge and allowed to parachute down to the surface, as they might easily do without risk.

Tame ducks are among the most sociable of birds and can even display bravery when threatened by a common enemy. The naturalist Houssay once learnt this as the result of a somewhat cruel experiment that he made in order to ascertain whether ducks invariably, [134] as alleged, fall upon a wounded comrade and destroy it. Wishing to satisfy himself on the point, Houssay, having come upon some ducks in a small pond, deliberately pelted them with stones till he had wounded one of their number. Instead, however, of behaving as he had been led to expect, the rest of the ducks formed close order round the wounded bird and sheltered it from further harm.

Few domestic animals—none, possibly, with the single exception of the camel—are less suggestive of "pets" than such gross poultry, yet even a gander, the most vicious tempered of them all, has been known to show lasting gratitude for an act of kindness. The bird, which had long been the terror of children in the little Devonshire village near which it lived, managed one day to get wedged in a drain, and there it would eventually have died unseen if a passing labourer had not seen its plight and set it at liberty. Down to the day of its death the bird, though nowise relinquishing its spiteful attitude towards others, followed its rustic benefactor about the place like a dog.

DECEMBER
THE ROBIN REDBREAST

[137]

THE ROBIN REDBREAST

Of all the old proverbs that are open to argument, few offer more material for criticism than that which has it that a good name is more easily lost than won; and if ever a living creature served to illustrate the converse to the proverbial dog with a bad name, that creature is the companionable little bird that we peculiarly associate with Christmas. Traditionally, the robin is a gentle little fellow of pious associations and with a tender fancy for covering the unburied dead with leaves; but in real life he is a little fire-eater, always ready to pick a quarrel with his less pugnacious neighbours. Yet so persistently does his good name cling, that, while ever ready to condemn the aggressive sparrow for the same fault, all of us have a good word for the robin, and in few of our wild birds are character and reputation so divergent.

Surely, however, the most interesting aspect of this familiar bird is its tameness, not to say attachment to ourselves, and so marked is its complete absence of fear that it is a wild bird in name only, and indeed [138] few cage birds are ever so bold as to perch on the gardener's spade on the look-out for the worms as he turns them up from the damp soil. The robin might, in fact, furnish the text of a lay-sermon on the fruits of kindness to animals, and those dialectical people who ask whether we are kind to the robin because it trusts us, or whether, on the other hand, it trusts us because we are kind to it, ask a foolish question that raises a wholly unnecessary confusion between cause and effect. It is a question that those, at any rate, who have seen the bird in countries where it is treated differently will have no difficulty whatever in answering. Broadly speaking, the redbreast has the best time of it in northern lands. This tolerance has not, as has been suggested, any connection with Protestantism, for such a distinction would exclude the greater part of Ireland, where, as it happens, the bird is as safe from persecution as in Britain, since the superstitious peasants firmly believe that anyone killing a "spiddog" will be punished by a lump growing on the palm of his hand. The untoward fate of the robin in Latin countries [139] bordering the Mediterranean has nothing to do with religion, but is merely the result of a pernicious habit of killing all manner of small birds for the table. The sight of rows of dead robins laid out on poulterers' stalls in the markets of Italy and southern France inspires such righteous indignation in British tourists as to make them forget for the moment that larks are exposed in the same way in Bond Street and at Leadenhall. In Italy and Provence, taught by sad experience the robin is as shy as any other small bird. It has learnt its lesson like the robins in the north, but the lesson is different. The most friendly robin I ever remember meeting with, out of England was in a garden attached to a café in Trebizond, where, hopping round my chair and picking up crumbs, it made me feel curiously at home. Similar treatment of other wild birds would in time produce the same result, and even the suspicious starling and stand-off rook might be taught to forget their fear of us. The robin, feeding less on fruit and grain than on worms and insects, has not made an enemy of the farmer or gardener. The common, too common, sparrow, [140] is another fearless neighbour, but its freedom from persecution, of late somewhat threatened by Sparrow Clubs, is due less to affection than to the futility of making any impression on such hordes as infest our streets.

No act of the robin's more forcibly illustrates its trust in man than the manner in which, at a season when all animals are abnormally shy and suspicious, it makes its nest not only near our dwellings, but actually in many cases under the same roof as ourselves. Letter-boxes, flowerpots, old boots, and bookshelves have all done duty, and I even remember a pair of robins, many years ago in Kent, bringing up two broods in an old rat trap which, fortunately too rusty to act, was still set and baited with a withered piece of bacon. Pages might be filled with the mere enumeration of curious and eccentric nesting sites chosen by this fearless bird, but a single proof of its indifference to the presence of man during the time of incubation may be cited from the MS. notebooks of the second Earl of Malmesbury, which I have read in the library at Heron Court. It seems that, while the east wing of that [141] pleasant mansion was being built, a pair of robins, having successfully brought up one family in one of the unfinished rooms, actually reared a second brood in a hole made for a scaffold-pole, though the sitting bird, being immediately beneath a plank on which the plasterers stood at work, was repeatedly splashed with mortar! The egg of the robin is subject to considerable variety of type. I think it was the late Lord

Lilford who, speaking on the subject of a Bill for the protection of wild birds' eggs, then before the House of Lords, gave it as his belief that no ornithologist of repute would swear to the name of a single British bird's egg without positively seeing one or other of the parent birds fly off the nest. This was, perhaps, a little overstating the difficulty of evidence, since any schoolboy with a fancy for birds-nesting might without hesitation identify such pronounced types as those of the chaffinch, with its purple blotches, the song-thrush with its black spots on a blue ground, or the nightingale, which resembles a miniature olive. Eggs, on the other hand, like those of the house sparrow, redshank [142] and some of the smaller warblers, are so easily confused with those of allied species that Lord Lilford's caution is by no means superfluous. Ordinarily speaking, the robin's egg is white, with red spots at one end, but I remember taking at Bexley, nearly thirty years ago, an immaculate one of coffee colour. As the robin is a favourite foster-parent with cuckoos, my first thought was that this might be an unusually small egg of the parasitic bird, which was very plentiful thereabouts. It so happened, however, that three days after I had abstracted the first and only egg I took from that nest, there was a second of the same type; and, much as I would have liked this also for my collection, I left it in the nest so as to set all doubts at rest. My moderation was rewarded, for no one else found the nest, and in due course the coffee-coloured egg produced a robin like the rest.

The robin is anything but a gregarious bird. Its fighting temper doubtless leads it to keep its own company, and we rarely see more than one singing on the same bush, or seeking for food on the same lawn. Yet, though it is with us all the year, it is known to perform [143] migrations within these islands, and possibly also overseas, chiefly connected with commissariat difficulties, and it is probable that on such occasions many robins may travel in company, though I have not been so fortunate as to come across them in their pilgrimage. Equally interesting, however, is the habit which the bird has in Devonshire of occasionally going down to the rocks on the seashore, as I have often noticed in the neighbourhood of Teignmouth and Torquay. What manner of food the redbreast may find in such surroundings is a mystery, but there it certainly spends some of its time, bobbing at the edge of the rock pools in much the same fashion as the dipper on inland waters.

Young robins are turned adrift at an early age to look after themselves, a result of the parent bird always rearing two families in the year, and in many cases even three, so that they have not too much time to devote to the upbringing of each. Another consequence of this prolific habit is that the robin has to make its nest earlier than most of our wild birds, and its nest has, in fact, been found near Torquay during the first week of January. [144]

It has long been the pardonable fancy of Englishmen exiled to new homes under the palms or pines, in the scorching tropical sun or in the biting northern blast, to misname all manner of conspicuous birds after well-remembered kinds left at home in the woods and fields of the old country. As might be expected of a bird so characteristic of English scenes, and so closely associated with the festival that always brings nostalgia to the emigrant, the robin has its share of these namesakes, and several of them bear little likeness to the original. In New South Wales, I remember being shown a "robin" which, though perhaps a little smaller, was not unlike our own bird, but the "robin" that was pointed out to me in the States, from Maine to Carolina, was as big as a thrush. Yet it had the red breast, by which, particularly conspicuous against a background of snow, this popular little bird is always recognisable, the male as well as the female. Indeed, to all outward appearance the sexes are absolutely alike, a striking contrast to the cock and hen pheasant, the first bird dealt with in these notes, as this is the last.

www.ingramcontent.com/pod-product-compliance
Lightning Source LLC
Chambersburg PA
CBHW030503220526
45464CB00006B/2643